What Do You Think About Ocean Animals?

DECIDE FOR YOURSELF

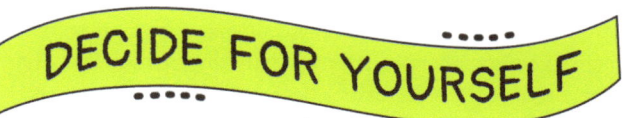

Kalena Baker

What Do You Think About Ocean Animals?
Copyright © 2025 by Kalena Baker. All rights reserved.
Published by Goose Creek, SC: Teaching Made Practical

No part of this work may be reproduced or transmitted in any form or by any means, electronic or mechanical, including photocopying and recording, or by any information storage or retrieval system, except as may be expressly permitted by the 1976 Copyright Act or in writing from the publisher.

Requests for permission or bulk orders, contact us at www.teachingmadepractical.com/books.
Visit our website at www.teachingmadepractical.com/books

Library of Congress Control Number: 2025920432
ISBN (hardback): 979-8-9915483-3-5
ISBN (paperback): 979-8-9915483-2-8

Editing by Deborah K. Frontiera, *authorsden.com/deborahkfrontiera*
Book design by Monica Thomas for TLC Book Design, *TLCBookDesign.com*
Images by BigStock, *www.bigstockphoto.com* and Adobe Stock, *stock.adobe.com*.

Publisher's Cataloging-in-Publication Data

Names: Baker, Kalena.
Title: What do you think about ocean animals? Decide for yourself / Kalena Baker.
Description: Goose Creek, SC : Teaching Made Practical, 2025. | Includes 47 color photos, 2 color diagrams, and 4 color maps. | Series: What do you think about ; book 2. | Audience: Ages 8-11. | Summary: Presents opposing opinions about five different ocean topics, each backed by selective facts. After reading the biased information, readers decide their own views on whether sharks are dangerous, jellyfish need a name, lionfish should be removed from their homes, fish farms are helpful or harmful, and what ocean animal is the smartest. Critical thinking questions are included.
Identifiers: LCCN 2025920432 | ISBN 9798991548335 (hardcover) | ISBN 9798991548328 (pbk.)
Subjects: LCSH: Marine animals – Juvenile literature. | Sharks – Juvenile literature. | Jellyfishes – Juvenile literature. | Fish culture – Juvenile literature. | Octopuses – Juvenile literature. | Dolphins – Juvenile literature. | Sea horses – Juvenile literature. | Critical thinking in children – Juvenile literature. | BISAC: JUVENILE NONFICTION / Animals / Marine Life. | JUVENILE NONFICTION / Science & Nature / Zoology. | JUVENILE NONFICTION / Games & Activities / Questions & Answers.
Classification: LCC QH541.5.S3 B35 2025 | J577.7 B--dc22
LC record available at https://lccn.loc.gov/2025920432

Table of Contents

About This Book 4

Sharks 6

Jellyfish 10

Lionfish 14

Fish Farms 18

Smartest Ocean Animal 22

Now It's Your Turn 26

Questions to Think About 28

Glossary 30

Index 31

About This Book

WARNING: This book is a little weird.

Like most ocean animal books, this book contains interesting animal facts. However, the facts are presented in an unusual way. Some facts will support one opinion while others will support the opposite viewpoint. Once you've read both sides of the argument, the decision is up to you. Which opinion do you think is right...or do you have a completely different opinion?

As you are reading, think about how the facts are being used to persuade you to think a certain way. What facts are emphasized? What facts are ignored? What facts are distorted? What words are being used to make you feel a certain way?

Remember, you don't have to agree with everything you read...even if the facts are true!

Shark Opinion #1

Sharks Are Dangerous

With their monstrous bodies, bone-crushing jaws, and rows of razor-sharp teeth, sharks are truly terrifying. Many species of sharks are **apex predators**, which means no other animal is big enough to hunt them. For example, the whale shark is the biggest fish in the ocean. It can grow over 40 feet long and weigh more than 40,000 pounds — that is more than 200 times heavier than an average human! How can something that big not be dangerous?

Protect Yourself

When you are at the beach, reduce your risk of a shark attack by:

- Staying with a buddy
- Staying near the shore
- Staying out of the water at dawn or dusk
- Not wearing shiny jewelry

Scientists believe that sharks do not intentionally hunt humans, but this doesn't make them less of a threat. Shark attacks still happen. When you are in the water, watch out — there is no outswimming a shark!

Great white sharks, tiger sharks, and bull sharks are known as "The Big 3" because they are the sharks most commonly involved in shark attacks on humans.

Shark Opinion #2

Sharks are NOT Dangerous

People get their absurd fear of sharks from TV and movies, not facts. The truth is sharks are mostly harmless fish that prefer to eat plankton, shrimp, fish, and crabs—not humans.

In 2024, there were only 88 shark bites worldwide, and just 9 deaths. Meanwhile, dogs bite *millions* of people every year, killing around 30–50 people. Even cows kill around 20 people each year! Movies should feature killer cows, not killer sharks.

At the beach, you don't need to worry about sharks. Sunburns, jellyfish stings, and rip currents are all much more hazardous.

? What do you think? Are sharks dangerous?

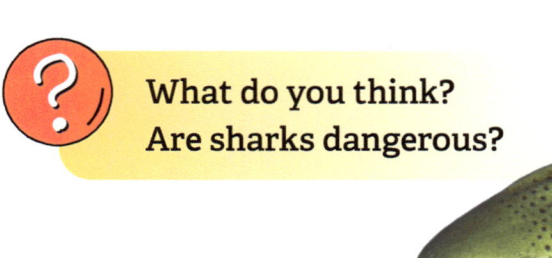

Zebra Shark

The largest shark in the ocean, the whale shark, is known as the "gentle giant" because it is completely harmless.

Scuba divers swim with sharks all the time without any problems!

Jellyfish Opinion #1

Jellyfish Need a New Name

The name "jellyfish" is incredibly misleading. Jellyfish look nothing like fish, because they aren't fish! They don't have gills, fins, scales, or backbones like real fish do. Jellyfish are actually cnidarians—**invertebrates** with stinging cells!

Long ago, people called anything that swam underwater a fish. But now, scientists know how to classify and name animals more accurately. That's why many scientists use the term "sea jellies" instead of jellyfish. Experts recognize the need for a name change, but sadly, most people ignorantly cling to the outdated name.

Did you know… a group of jellyfish is called a smack?

Jellyfish body parts are completely different from fish body parts!

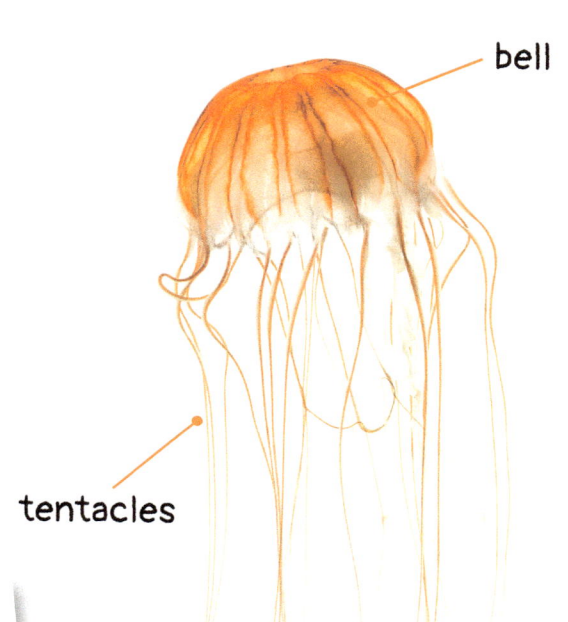

bell

tentacles

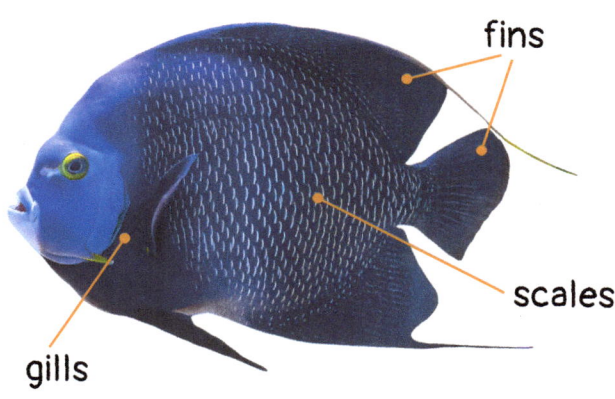

fins

scales

gills

Jellyfish Opinion #2

Jellyfish Do NOT Need a New Name

A handful of people are fighting for jellyfish to get a new name—sea jelly. That might be more precise, but jellyfish is a fun, easy to remember name that people have been using for centuries.

There are all sorts of words in the English language that are imprecise. Greenland isn't really green—it's mostly ice! Guinea pigs aren't really pigs, and they aren't from Guinea. The sun doesn't really rise, yet we still watch sunrises. Imagine how confusing it would be to update ALL of the imprecise words in our language. Why start with jellyfish?

What do you think? Do jellyfish need a new name?

Crayfish

Silverfish

Starfish

Cuttlefish

Jellyfish aren't the only non-fish with fish in their names!

Lionfish Opinion #1

Lionfish Should Be Removed from Their Homes

Lionfish are invading the Atlantic Ocean. The first lionfish, probably former pets, were spotted off the coast of Florida in the 1980s. A few **invasive** animals were not a problem, but female lionfish can lay 2 million eggs every year, and they have no natural predators to stop them. Now, the Atlantic Ocean is overrun with these destructive creatures.

If lionfish are not removed from the Atlantic Ocean, then entire ecosystems will be destroyed. They steal homes and food from **native** fish and eat the fish that keep coral reefs healthy. As lionfish numbers grow, other fish species start to disappear. To protect ocean life, we must get rid of the lionfish before it's too late.

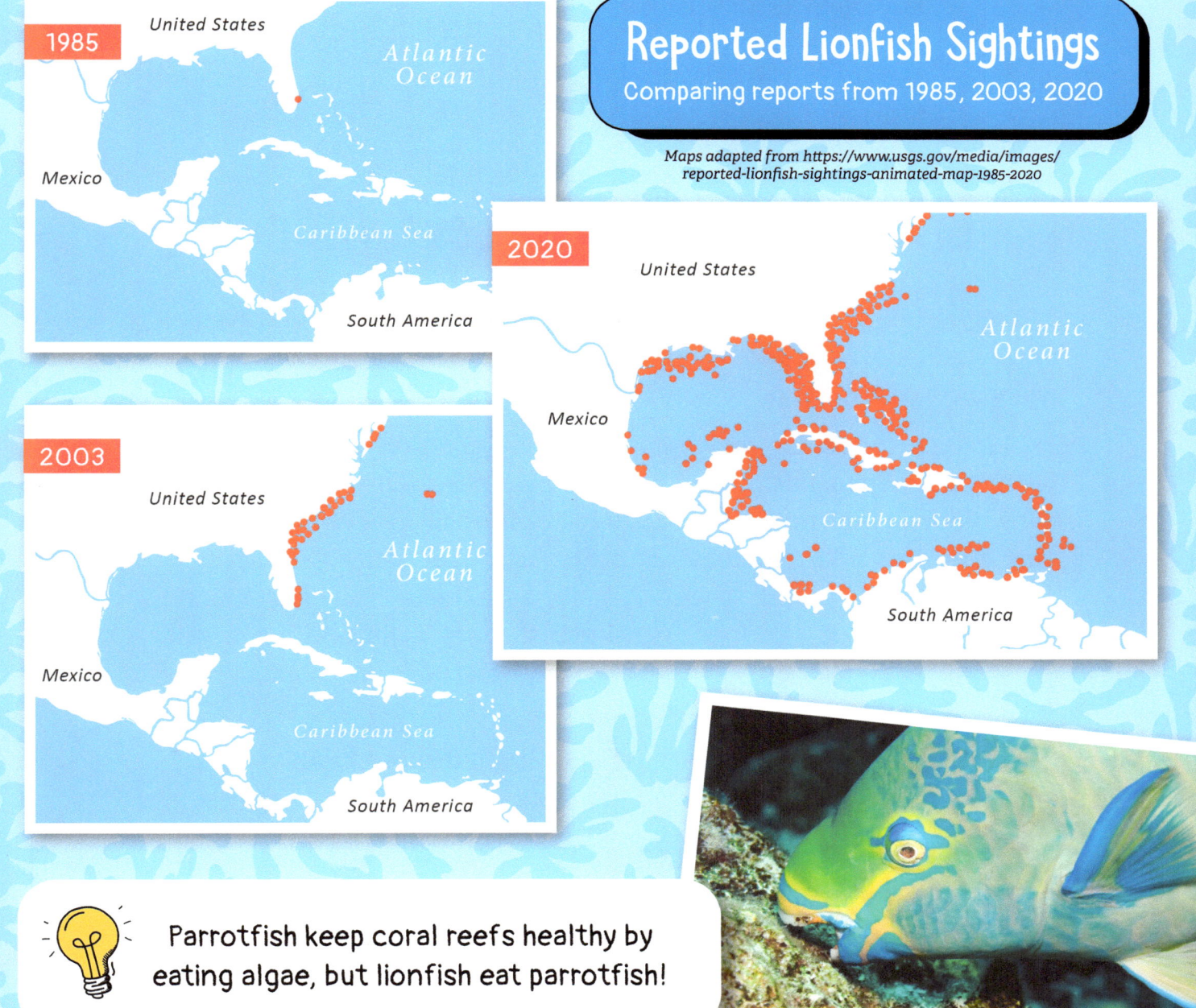

Reported Lionfish Sightings
Comparing reports from 1985, 2003, 2020

Maps adapted from https://www.usgs.gov/media/images/reported-lionfish-sightings-animated-map-1985-2020

Parrotfish keep coral reefs healthy by eating algae, but lionfish eat parrotfish!

Lionfish Opinion #2

Lionfish Should Be Left Alone

Imagine being kicked out of your home just because someone decided you didn't belong. That would be despicable! But that is exactly what some people want to do to lionfish in the Atlantic Ocean.

Lionfish didn't migrate to the Atlantic Ocean on their own. Most likely, an irresponsible human dumped their pet lionfish into the ocean. Now, lionfish are simply trying to survive in their new homes, just like any animal would. Yes, they eat other fish—but that's what fish do! They have never killed a human. Lionfish have done nothing wrong, so they deserve to be left alone!

? What do you think? Should lionfish be removed from their homes?

Lionfish are **native** to the Indian and Pacific Oceans, where they live peacefully and in balance with other sea life.

Native Habitat of Lionfish

Fish Farm Opinion #1

Fish Farms Are Helpful

The ocean is running out of fish! **Overfishing**—catching too many fish too quickly—has **endangered** many fish species. This is a massive problem because millions of people rely on fish for food. Fish farming is the only way to feed everyone without wiping out wild fish populations.

On fish farms, people known as **aquaculturists** raise fish in tanks or pens, just like farmers raise cows or grow corn. This helps keep wild fish in the ocean where they belong. Fish farms also provide people with steady jobs—while making the cost of fish cheaper!

So, what's not to love? Fish farming helps feed hungry people at an affordable price, protects wild fish, and provides jobs. It's a win-win-win!

> **Did you know...**
> over 3 billion people rely on seafood to live?

Overfished Ocean Animals

Sea cucumbers are a delicacy in some countries, and because of this, are **endangered**.

Sturgeon are **endangered** because people like to eat their eggs, known as caviar.

Giant manta rays are **overfished** by people who want to use their gills in medicine.

Fish Farm Opinion #2

Fish Farms Are Harmful

Fish farming sounds like a harmless way to raise fish, but it's not—it's destructive.

Fish farms cram a LOT of fish into crowded spaces, making it easy for disease to spread like wildfire. Farmers use **antibiotics** to fight these diseases, but the antibiotics often leak into the ocean, destroying nearby wildlife and impacting the people who eat the fish!

It gets worse. All those fish produce a ton of waste, polluting the water. Like diseases, water **pollution** spreads easily to nearby oceans and rivers. And once these ocean environments are destroyed, they are gone forever.

Pollution and diseases from fish farms easily spread to the nearby ocean environments.

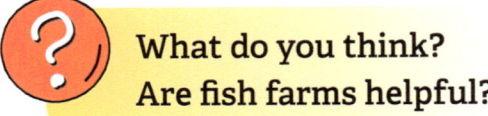

What do you think? Are fish farms helpful?

Wild fish have plenty of space to roam, but fish farms crowd fish into small spaces.

Wild Fish

Farmed Fish

Smartest Ocean Animal Opinion #1

Octopuses Are the Smartest Ocean Animal

If you're looking for the smartest animal in the ocean, then the obvious winner is the animal that has not one, but NINE brains—the octopus! Each of an octopus's eight arms has its own mini-brain, on top of its main central brain.

Octopus brains are also bigger than any other **invertebrate** in proportion to their size! These powerful brains help octopuses solve problems, use tools, and escape mazes. And unlike many other animals, octopuses don't need any help from mom or dad. Baby octopuses are so clever that they figure everything out on their own!

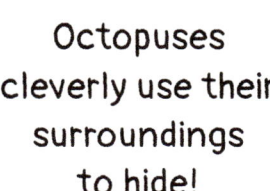
Octopuses cleverly use their surroundings to hide!

Smartest Ocean Animal Opinion #2

Dolphins Are the Smartest Ocean Animal

Dolphins aren't your ordinary ocean animal—they're the geniuses of the sea! From birth, dolphins "talk" using whistles, squeaks, and clicks. Each dolphin even has its own signature whistle, like a name. Dolphins can recognize each other using that whistle, even after years apart. That's something most animals could never do!

> **Did you know…**
> Dolphins have the biggest brains for their size of any animal *except* humans!

Dolphins are so smart that scientists taught some of them to recognize symbols, just like humans learn letters. Dolphins can read! But that's not all. Dolphins use tools, solve problems, play games, and show kindness. What other ocean animal can do all these clever things?

 **What do you think is the smartest ocean animal?**

Dolphins can learn to perform a variety of tricks because of their ability to adapt and solve problems.

Dolphins work in groups to hunt, smartly surrounding schools of fish and taking turns eating.

Now It's Your Turn

Use the seahorse facts below to write 2 different opinions:

1. Seahorses are terrible predators.
2. Seahorses are excellent predators.

Seahorse Facts

- Seahorses have uniquely shaped heads that let them sneak up on prey without creating waves in the water.
- Seahorses are tiny — the smallest seahorse species is only about ½ an inch tall.
- Seahorses are known as the slowest fish in the ocean; the dwarf seahorse has a top speed of about five feet per hour.
- Seahorses have long snouts that slurp up prey like a vacuum.
- Seahorses are always on the hunt, eating 30 to 50 times a day.
- Because seahorses have no teeth, they cannot bite or chew their prey.

Seahorses are well camouflaged, making it easy for them to stealthily surprise their prey.

 Questions to Think About

Sharks

- Both passages used whale shark facts to support their opinion. What facts about whale sharks did each passage emphasize? Which facts are most relevant?
- What words did the first passage use to make sharks sound more dangerous? What words did the second passage use to make sharks sound less dangerous?

Jellyfish

- How did the first passage insult those who continued to use the name jellyfish? Is this insult a persuasive argument?
- Both passages agree that sea jelly is a more accurate name for jellyfish. Why do they come to different conclusions?

Lionfish

- The first passage focuses on the problems lionfish are creating, while the second passage focuses on the harm being done to lionfish. Which do you think is more important?
- What does the second passage say about how to fix the problems lionfish are causing? What does this tell you about the author?

Questions to Think About

Fish Farms

- What problems does the first passage leave out? What problems does the second passage leave out?

- What do you think is more important — feeding people cheaply, like the first passage emphasizes, or protecting the ocean environment, like the second passage emphasizes?

Smartest Ocean Animal

- According to the first passage, what makes an animal smart? What makes an animal smart according to the second passage? What do you think makes an animal smart?

- Both passages talk about the size of the animal brain. What additional information would be helpful for comparing octopus and dolphin brains?

Remember, you do not have to agree with every opinion you read — even if the facts are true!

 Glossary

Antibiotics—medicines that help fight infections

Apex predators—animals at the top of the food chain that other animals don't hunt

Aquaculturist—a person who farms ocean animals or plants

Endangered—when something is in danger of disappearing forever

Invasive species—a plant or animal that moves to a new place and causes problems for the plants and animals that already live there

Invertebrate—an animal without a backbone

Native species—a plant or animal that naturally lives in a certain area

Overfishing—catching too many fish too fast, before new fish can be born to replace them

Pollution—trash or chemicals that make the earth dirty

Predator—an animal that hunts and eats other animals

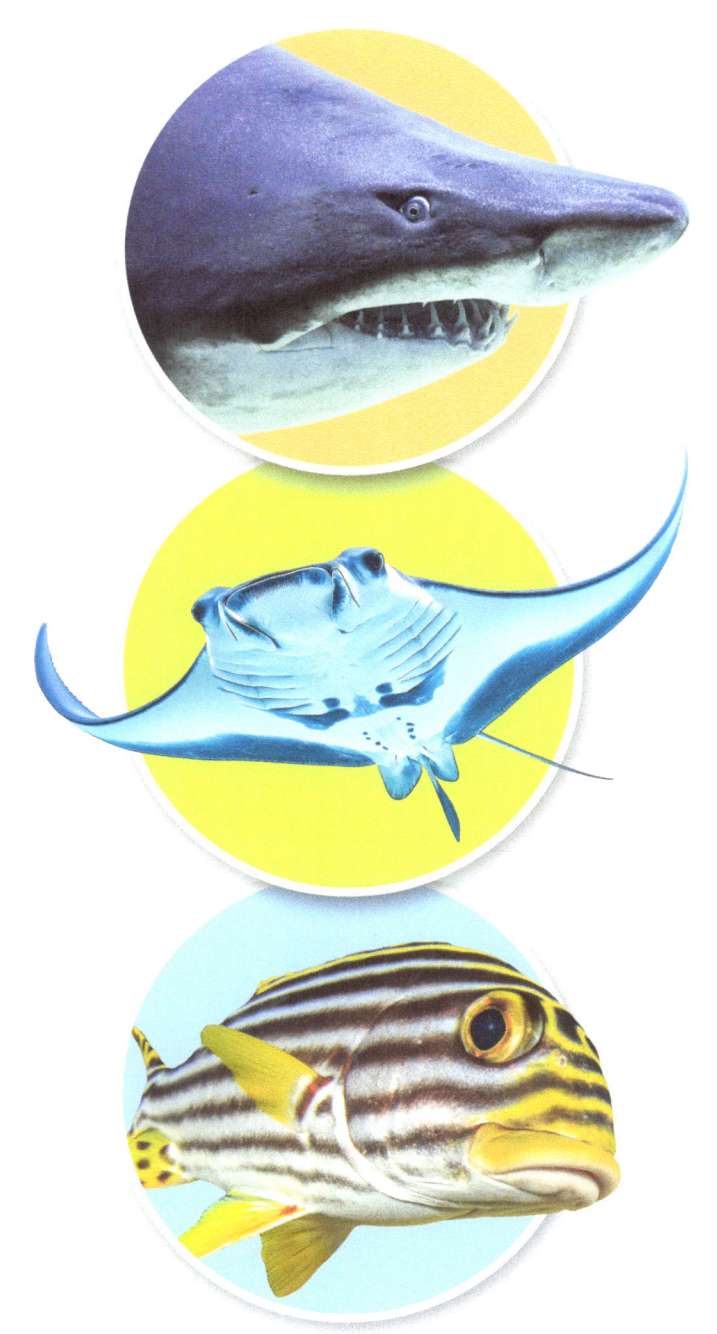

Index

Dolphin, 24–25, 29

Fish Farm, 18–21, 29

Invertebrate, 10, 22, 30

Jellyfish, 8, 10–13, 28

Lionfish, 14–17, 28

Manta ray, 19

Octopus, 22–23, 29

Seahorse, 26–27

Shark, 6–9, 28

Starfish, 13

Parents & Teachers

Get free printables to use with this book at www.teachingmadepractical.com/whatdoyouthinkaboutoceananimals.

www.ingramcontent.com/pod-product-compliance
Lightning Source LLC
Chambersburg PA
CBHW051515110526
44582CB00007B/130